# Moving on Through

# Moving on Through

POETRY BY TOM PAYNE

# PALMETTO
## PUBLISHING
Charleston, SC
www.PalmettoPublishing.com

Paperback ISBN: 979-8-8229-2982-1

# They bless my life

*My mother Beth: surly an angel*

*My son Cody: every day my sun*

*My wife Jacquie: everywhere a garden, everywhere love*

*My dearest friend Barney: so many versions of great heart*

# WHAT I THINK

I think the black dog in the Dragon Bar
   last night
Looked me in the eye and said,
"Badger says hello."

I think every time I see a bird
My mother sees it too
And that blessings are as simple as a pond
And a trout and a "Dad, I caught one."

I think in every smile there is grace,
That heaven sounds like frog song,
That peace comes often but well disguised.
I think Shakespeare was right when he said
"There is more on heaven and earth
Than your science can account for."

I think when we left Montana that day
And heard the "Last Train Home,"
We knew what was to come.

I think there are phalanxes of muskox
And barren ground caribou
And Arctics that I will never see.
But they are there and that's wonderful.
And I think that when I say "thank you
That's who I want to be.

# PANAMINT RANGE

Sometimes things don't work just right,
A busted pickup,
Someone dying on you.

Maybe there's too much room out there,
Horizons stretching dreams beyond ridge after ridge
Only, years later, to turn them
Into the coyote howls in coulees.

It's a place of try and failure,
A place of river bottoms and basins
That tilt up into mountain passes that cut right through the
  sky.
A place where you're glad that at least you were there
When things went wrong.

# CODY IN NEW MEXICO

Coming out of the hills at dusk,
Down from Silver City, New Mexico,
Ranges stretch forever in the distance
In hundreds of serrated nuances of purple:
The Chiricahuas, the Mogollons, the Gilas, the Giliuros.
There is nothing else.
For a moment it is a frontier,
The edge of something vast and wondrous.
My son is in the back seat, asleep.
"Look," I want to say to him, "There is nothing out here
Except darkening sky and mountains.
Here is where the coyote lives,
And the badger and the fox."

He is three years old and I won't tell him
How far we've had to come to be here,
I won't tell him how much I want him to remember
Being in this place with me.

A lesser nighthawk swerves across the sky
And soon the ocotillo darkens into silhouette:
Lordsburg, a dusty glow on the horizon,
Breaks the vision.
I turn to look at him and whisper,
"Come back to this place with me, Cody:
And when you can't come back with me,
Come back."

# THE TREE AND THE SOUND OF THE DOOR

It stuck when he tried to close it and he remembered to fix it,
Just plane the bottom a bit.
And then he remembered that he would never open
    or close it again.

It shut. And kept shutting forever after.
In the middle of what joy he had, of moments of hope,
He could hear the door shut that final time:
A sad, big something closing behind him.

The tree was different, though. He liked remembering it,
The cutting into the bank, the shovel digging it out and,
    home, the digging it back in,
A stick with two branches, wires and braces to hold it up.
And as quiet as the yearly pulse of frog song,
Of years of the comings and goings in the sky,
It grew. Big. The only one out there.

In the winter its branches moaned like masts of ships,
And in the spring it snowed flurries of cotton tufts,
Lining for the nests of thrush and finch.
In the bake of summers it had whispered the opiate of breezes.

The tree would always be different because when try ran out,
It remained,
Something worth the effort after all.

# RAPHAEL IN OKLAHOMA

Left the terrors behind and hit the road.
Ended up in a short grass corner of Colorado and Oklahoma.
Went through 23 iPod tunes. Beautiful, good day.

Had a BLT somewhere
At a three stool counter in the only place in town.
The young waitress, a painting by Raphael.
She glowed, suspended somewhere beyond sorrow and time.
Frog tattoo on her shoulder.

I said, like a prayer, "Hi."
 She smiled a "How are you?" like she meant it.
And I didn't say, "I don't know anymore.
It doesn't get simpler does it?
But I'm working on it."
 She wouldn't know about the troubles.
But I did say, "Great day, still a lot of fires, though,"
And she said, "Well, She knows what She's doing."
And I thought, "I don't know who She is but if I did,
     would I get to glow, too?
Can I? Probably not anymore. How did I miss that?"

When I finished, left a big tip and said,
"Thanks for your radiance" and meant it.
Headed into a dark sky over the Rockies.

An hour later saw a huge antelope curled up, legs folded,
    on the prairie.
He owned it. He glowed.
He knew who She was too.

# SOUTHWEST SUMMER NIGHT

Dirt road to the Silver Lounge,
Saturday night band with steel guitar,
Pressed-out shirts and boots, dusted off.

No moon.  Rutted miles along fence and brush.
Windows down, still hot,
Dust and jack rabbits in the headlights,
Snap of nighthawks swerving by.
Last three cans of Bud getting warm.
Stiff-legged javelina in a draw,
Lordsberg tunes fade in and out.

Hour later County Road 17,
Left, thirty miles of cholla to the interstate,
The La Grande ranch and beyond,
To lightening outlines of the Chiricahuas.
Soon flooding in the canyons up above.

Right, just two miles and in a draw
Old pickups parked in ocotillo scrub.
Adobe walls with bare-bulb lights strung on top,
Noise and smoke leaking out.
Two hands throwing up, more fights inside.
Silver Lounge.

Johnny Horse already drunk. Big grin.  Last tooth gone,
    another heifer horn.

Crystal broke up, trucker from Las Cruces,
Half pint in her purse to get her home.

Jamie who never kept a dime, son now dead,
     her kidney going bad.
Bobbie, rode the last two miles on the rims,
     drinking Jager shots.

Free chips and ketchup, drummer setting up his snares,
Zig Zags and Bugle and Benny's longhorn buckle,
Gold-plated, passed around.

Juke box plays vinyl sorrow songs
About drinkers and loners and hands and try:
About Mother Lodes and a Silver Lounge.

# THE CANADIAN PACIFIC ACROSS MANITOBA

By cupping my hands I can see through reflection, outside,
Where track-side fence posts blur by in Pullman
    rectangles of light
And black outlines of ridges spiked with conifers
    which undulate in silhouette.
An hour later the view flattens out into endless fields,
To miles of fence-post, fence-post, fence-post in the night,
To an empty snowy nothing else.

Nothing gives relief, nothing breaks the distance
Until a single light, like a single star in a black sky,
Sputters across the field into view.
 It bobs in the windy night, steadies,
And, as the train nears, becomes the the single streetlight
At the single intersection of a tiny town.
It drops a yellow circle on two, car-less, empty streets,
Illuminates sockets of windows and leafless trees.

The train moves on through and I press against the glass
But it does no good: It all speeds past...
    The little town,
    The house, the white fence, the loose board on the porch,
    The lilacs, the morning sunlight on the kitchen table,
    The dog in the shade of the red barn, the tire swing,
    An entire field of meadowlarks,
    My mother baking bread.
It all drops back and disappears into the prairie night.

# BEARS DEN KARAOKE

Thursday nights at 8:00,
Two dollar Buds.
Moose and deer and bears stare down
In this Maine woods cocktail lounge.

Chairs moved against the walls,
Table and microphone up in front

Now people wait their turn.
Been rehearsing below-zero months to the radio,
Listening for the song just right,
The one that might tell it all:
Tell when the heating oil ran out,
When someone was suddenly left behind,
When the free food shelf in town was empty.
For the one that might tell it all,
To someone who might want to listen.

When their turn is called, they move up front,
Anxious that this too might go wrong.
But their songs are wonderful, off-key arias.
They sing because they need to,
Because they have a song.

And in this Maine woods cocktail lounge
Their songs are like the tales the Elders told
Around fires long ago.

# MY FATHER'S SHOES

Next to the Hotel Utah, "The Gentleman's Haberdashery:
Tailored Suits, Accessories, the Finest Shirts,
    Men's Dress Shoes."

He bought Florsheims, the best they had.
In the box each in its own felt sleeve
And when he withdrew them they shone,
Looked in his hands like little wingtip yachts.

Soon he was walking down South Main Street
Their thick leather soles drumming on the pavement
    like a Scottish Tattoo.

Sixty years later they were the only shoes he had.
But he polished them almost every day
So the leather, now cracked and split in places,
Glowed like the patina of old mahogany.

When special had run out
These still were.

When he died there was nothing left
But the shoes, waiting again to go outside.
They were heavy and smelled liked wax,
The grain something old and treasured.

The soles, though, told more about him
Than I ever knew:
They had huge holes all the way through.

No one would have ever known.
To my father they could always be those
    shinny little yachts,
Their leaks invisible.

To me the Florsheim's looked like dignity.

# OREGON
# (DROPPING CODY OFF
# TO HIS NEW JOB)

The Snake, the Umpqua, the Mackenzie, the Willamite,
    the Deschutes,  the Green River,
They all run through.

Green River, a town bypassed by a highway.
"Well, only two restaurants left to choose from
But great steaks, still, at Roy's."
Outfitters and pizza parlors and all else closed.
Railroad sounds at night.

Twin Falls, the Snake runs past.
John Deere dealers and fields irrigated by dinosaur spines.
Bend on the Deschutes.

Eugene on the Willamite.
Left in 1965, left much more in 2016.
An end and a beginning in the green and rain.

# TOWN TRACKS

This is where he came from:
A silo, train tracks, a bar,
A feed store,
A wind as precise in naming months as any calendar.

It was the tracks, though,
The big Santa Fe's pulling right through,
Straight West, Straight East,
Year after year, that pulled him away
To some unknown but certain salvation.

And today, tracks that splinter into the rail yards of
Those golden places once far away,
Now pull him back, perhaps to rescue.

# LATE NIGHT: DOVER-FOXCROFT BAR

Each night he waits until the bar empties.
Now it's just the woman who works there,
   busy cleaning things.
He takes out five ones
And feeds them into the juke box.
Knows just what to play...
It's the same tunes every night:
J15, W51, A14 and A15, H61....
Until the entire five is gone.

I asked him one night
(I was late to leave),
"Why the same songs?"
He said, "They were my dad's favorites,
They were things he  gave me."

# THE FOLKSTON FUNNEL

On the way, down towards the Okeefenokee,
Past the Squat and Gobble Cafe,
Hundreds of Divine Worship churches,
Scrawny pine woods and trailer parks or trailers
    not in parks
That collapse inwards, single lights in their windows
    as if for rescue.

Past Hardees and Dollar Stores and, under curtains
    of Spanish moss,
Yard sales,  card tables of scraps of what's left over
    from not much to begin with.
Past sagging porches disappearing into kudzu
And everywhere fields of parts of things.

On Highway 1, Folkston,
Where people wait in pavilions to watch trains,
    60 per day and night, go by.
The Folkston Funnel, the clot in the artery of rail lines
    North to South, South to North.
Today a gathering to watch the 5:32 from Jacksonville.
And, yes, at 5:32, that's the one from Jacksonville.
    Right on time.

# FOR ROCKY

Old dog stretches its tired legs
Towards the crackling fire
And goes back to sleep.

I glance up, smile,
And wait for him to stir again.

# FOR BADGER

He sits on the dock
Looking at a sunset across the glass-smooth lake.
He doesn't move, transfixed by the red and orange
That shift and finally dissolve into gray outline of ridge.

His stare is unbroken
In an uninterrupted vision of sky
And green hemispheres faraway.

I think, "So that's what peace looks like"
And thank him for the lesson.

# SAN ISIDOR, ECUADOR

On the gray wall of a wooden shed,
Moths.
Hand-sized mimics of leaves and mosses
in the myriad jagged patterns of the forest.
I name them—-
Cecropia, Io, Silk, Hawk—-
And need to touch one,
To see its exquisite soft wings spread across my palm.

When I try, it flies,
Struggles up high enough for the sudden vector of a falcon
To take it and disappear away.

Just because I touched it,
Just because I couldn't leave this powdered wonder alone.

# CAMP POEMS

# THE GRAYING SKY

Gray means corn and apples and chanterelles,
A jacket against the cold, a fire inside,
And a turning within just as the weather turns.

Gray means migration, flocks moving over the lake,
Ferns that curl up and disappear into the soil,
A time for a counting of passing seasons.

For those we love
Let gray be just precursor of another spring,
Its phlox and beans and frog sound,
The comfort of its sun.

Let them see gray as another kind of light.

# MAINE CAMP CLOSING

And then it grows quiet.
Boats gone, docks dragged into the woods, doors locked.
Now just a lake, maples turning, some last birds at the feeder.
And a breeze from the North for a while,
Warm like a spring day long ago.
And there are memories in it and the caress of well being.
Fragile things, balanced in a late season,
Calling like the voice of loons.

# A MOMENT

Mink on the dock, loons behind,
Then both gone.
So suddenly wonders come and go.

# FUGUE

Now a chorus without frog song,
But owls and loons and crows still call,
And black-eyed susans burst out loud.

Thus voices watch the conductor
Move the baton of seasons across the sky:
Wait and wait and wait
Until each gets to sing again.

# JO MARY WILDERNESS

Radio stations mostly French up here
And cars have canoes on top.
At night on the ponds
Moose and loons and trout.
And by casting out a fly I lasso the stars.

# WATER SKATERS

Dusk.  A quarter moon.
The lake smooths out to inky glass.
A curtain parts, pulled by some invisible, circadian hand.
And thousands of skaters take the stage.
They spin and dart across the water,
Ions of insect energy, streaks of life.
Like ours, each in performance of a lifetime,
Each a one-season dance.

# FLEDGLINGS

We know about fledglings, about feathers
growing out of shafts,
About their balance on the edge of nests.
And, easiest, we know about their wide-gaped bills,
The frenzied fluttering of downy wings,
Their incessant chirping, more and more, for meals.
And we know that suddenly they are gone.

What we don't know is the afterwards:
How the parents continue to search branch to branch,
Looking for something lost, something they need to feed.

# ZUGUNRUE

# CENTRAL VALLEY (FROM ZUGUNRUE)

It was late afternoon so he decided to take his favorite drive again, up Highway 5, and have a drink at Louie Cairos, sixty miles out of Sacramento.  Late summer and the harvest smell of Central Valley fields---acres of milo and tomatoes, oats and alfalfa---reached as far back as memory could.  Ground squirrels sat on top of fence posts to escape the heat of their burrows; pheasants exploded out of irrigation ditches where egrets waded to catch frogs; red-winged blackbirds, their scarlet patches almost gone, balanced from the tops of cattails.

All these things, so familiar for so many years.  And then, just before he pulled off the interstate into Williams, a caravan of trucks passed, towing carnival rides from the California State Fair that had just ended.  Gayly painted trailers carrying  Corn Dog stands, Tilt-a-Wheels, Gyros booths, Zippers, SnoCone concessions, World's Largest Frog and Smallest Horse tents,  girders of roller coasters and Ferris wheels rolled by, and it was like watching his summer, all his summers, disappearing over the edge of the world.

He was detached from work, from people, from his house, what was left?  Now there was not even the possibility of a deep-fried Twinkie or a ride on the Super Lumberjack Log Jamb.

But what he expected in Williams he happily got.  A place still connected to the earth; people who still got dirt on their shoes and waited for weather or waited out weather; who could hear the seasons fly over in the sky.  A town with a railroad track down the middle and a grain elevator skyline.  A good town.

Driving through. Simple things. A farm. A fat man on a bench. A dog in the shade. A red barn. A white fence. A frame house. Clothes on a line. A porch and screen door with no screen. A crow on a plow disc. A tire swing. A worn broom. A horse. A green tarp on top of a bale of hay. In back of it all, a dirt road West, up through oaks and hills and pines, then down to fog and coast.

Benny's Guns, the Duck Inn, Jim's Feed Store, the Blue Cafe: a main street of decent, unchanging places.

Louie Cairo himself was behind the bar, been there for fifty years. His genus dog Buggs sat on a stool at the bar happily chewing a smoked sausage stick,his head even with the other drinkers, who had probably been there out of the heat for hours.

Behind them, two tables of Mexicans sat around drinking pitchers of beer. They were covered with the dirt from working a long harvest day, and they were laughing and happy as if the season of hard work was done and they didn't have to get up at 5:00 the next morning and the next mornings of the rest of their lives and work with their hands and sore backs. They were happy because right now, they could drink beer and laugh with each other. And that was enough.

A flabby woman with dirty gray hair and a kind, shy smile came in with a pie. She was wearing worn flip-flops and a cotton dress, thread-bear almost to the point of transparency, washed beyond any recognizable pattern or form. The Mexicans called her "Tia," and she said hello to them in Spanish and bought Buggs another sausage stick and took the pie in back. Karen the hostess came in to say the kitchen was open but Mary, the server, had called in sick because her dad had fallen again and she had to stay home with him. And the flabby woman and Louie the owner of Buggs the genus dog and everyone at the bar and the Mexicans agreed that was bad and they put money in a 49'ers cap to help her out. Again.

Someone walked to jukebox and pretty soon Louie Carios was filled with "Lukenbach Texas" and the place was filled with everybody...Louie, the bar crowd, the Mexicans, Tia, Karen, even Buggs...singing along.

He pealed the label of his second bottle of beer and when the lyrics got to "Let's get back to the basics of life," it sounded like the voice that haunts his nights instead of dreams.

He finised his beer and stepped outside, blinded by the heat and sun. Just as was he getting into his car, a dent-and-Bondo pick-up rolled across the gravel parking lot. In the back was a 30-pound carp Ernie had caught with a dough ball in "the slough." It looked prehistoric, bronze scaled, something that had endured time. Something from the mud. Ernie was pretty sure he could smoke it, "get rid of its stink." Buggs the genus dog came out to smel it and then ran back inside and jumped up on his stool.

As he drove away, he thought about Ernie's stinky fish and Buggs the genus dog and Tia's pies and the pitchers of beer and the ball cap donations and the laughter, and he realized he was smiing.

An hour later back on Highway 5, a short-eared owl flew stiff-winged across the road and harvesters moved slowly across the fields, leaving in the evening light illuminated trails of dust which rose and then disintegrated like memories.

# A DRIVE TO WORK (FROM ZUGUNRUE)

He was late the next morning but not from the predictable hangover, which predictably allowed no self pity but only a surfeit of self-reproach. He knew he could ride it out. No, he was late from yet another wonderful drive to work.

Every morning began with a wonderful drive to work. All he had to do was look around. This morning on 43rd and M Street, there were two crows meticulously sorting through the contents of a Burger King bag, a corvidae Mother Lode, discarded in the gutter. Then, Alhambra and J Street, a SMUD worker disappearing down a manhole, penetrating the peel of the earth in his own way. A long way to go to the core. On 25th and K, the tattooed vixen in pajamas pressure washing the sidewalk in front of the Press Club, its marquee announcing that tonight the "Snobs" would be playing. A block later, a woman dragging her Dalmatian, who knew exactly where he was going into the Mid-City Veterinarians. Then the free grocery line, Lutheran Church, elderly immigrants mostly, having learned lessons of endurance from places faraway. A wasp nest in the sycamore in front of David's Pawn. Next block, the fury of a man pumping up a bicycle tire, desperate strokes up and down. A canopy of elms, the cumulus of lawn sprinklers, three men waiting at the Zebra Barber Shop, "Five dollar special before 9:00 AM."

And the best thing of all, 19th Street, a freight train heading East, towards the Sierra, the Great Basin and the Great Salt Lake, past Aunt Mabel picking blackberries in the Wasatch Mountains, past the Ute Reservation homestead where his mom was born, past buffalo runs, past brown bears plunging into a salmon spawn, past abandoned farms whose broken windows stare out like the blind, past giant cottonwoods and river bottoms and the patches of snow on mountain passes.

Into and over the Rockies--- stopping traffic for fourteen minutes, four Santa Fe locomotives, then gondolas, tank cars, flat cars with rolls of steel strapped down, ro-ro freight cars, refrigerator cars, all with wonderful names: Lackawanna, Southern Pacific, Wabash, Lake Erie Rail Road Company, Northern Ohio, Penn Central, Reading, Union Pacific. And then he added some others he knew: the Aurora Borealis Express, the Glacier Express, the Zephyr.

It was like the entire lower forty-eight rolling by, on display. And when the semaphore rose and the flashing lights stopped and there was no caboose (he had imagined for years brakemen in cabooses, smoking pipes and reading Tennyson by the little pot-bellied stoves at the end of trains crossing the Dakotas in sub-zero winter nights) and traffic moved forward again, it was anticlimax. Nothing left. But to go to work.

# GREYHOUND
# (FROM ZUGUNRUE)

`The best part of the bad part of town was the cavernous shed of the Greyhound Bus Depot with its lines of people waiting in the diesel fumes for the doors to open into buses marked "New York," "Los Angeles," "Reno," "Miami," "Bakersfield," "Tijuana."

Before the drivers in grey uniforms arrived to punch tickets and board passengers, the travelers smoked or retied the strings that held their boxes together or finished a donut or just stood dazed, humbled by exhaustion and hope.

Then there were always the forgotten, usually men, wearing worn suit coats and carrying, incongruously, those old, square women's cosmetic cases, which held the Joseph Cornell collections of their lives (a creased photograph of a woman and boy eating ice cream; a small, ceramic penguin; a pencil list of names; a faded, pale-blue letter; a match book cover from the Hotel Utah). They squint at the destinations displayed on the front of buses, puzzling, perhaps, over the choice for a final journey.

This was "migration," too: the process of life in perpetual movement towards the dream of a happy place. Zungunrue. As he always did in front of the Greyhound Bus Depot, Tom bowed to the scene with reverence and whispered the four T.S. Eliot lines:

> *"I am moved by fancies that are curled*
> *Around these images, and cling:*
> *The notion of some infinitely gentle*
> *Infinitely suffering thing."*

Milton Keynes UK
Ingram Content Group UK Ltd.
UKHW050737120224
437693UK00007B/95

9 798822 929821